RETURN FROM EXTINCTION

THE GRAY WOLF

LEONARD CLASKY

Gareth Stevens PUBLISHING

Please visit our website, www.garethstevens.com. For a free color catalog of all our high-quality books, call toll free 1-800-542-2595 or fax 1-877-542-2596.

Portions of this work were originally authored by Caitie McAneney and published as *The Return of the Gray Wolf.* All new material this edition authored by Leonard Clasky.

Library of Congress Cataloging-in-Publication Data
Names: Clasky, Leonard, author.
Title: The gray wolf / Leonard Clasky.
Description: New York : Gareth Stevens Publishing, [2023] | Series: Return from extinction | Includes index.
Identifiers: LCCN 2021062261 (print) | LCCN 2021062262 (ebook) | ISBN 9781538280294 (library binding) | ISBN 9781538280270 (paperback) | ISBN 9781538280287 (set) | ISBN 9781538280300 (ebook)
Subjects: LCSH: Gray wolf–Conservation–Yellowstone National Park–Juvenile literature. | Gray wolf–Reintroduction–Yellowstone National Park–Juvenile literature. | Yellowstone National Park–Juvenile literature.
Classification: LCC QL737.C22 C585 2023 (print) | LCC QL737.C22 (ebook) | DDC 599.77309787/52–dc23/eng/20220124
LC record available at https://lccn.loc.gov/2021062261
LC ebook record available at https://lccn.loc.gov/2021062262
Published in 2023 by
Gareth Stevens Publishing
29 E. 21st Street
New York, NY 10010

Designer: Olivia Lo Sardo
Editor: Caitie McAneney

Photo credits: Cover MartinMencarowski/Shutterstock; p5 HollyKuchera/Shutterstock; p7 RaphaelRivest/Shutterstock; p8 StoriesInLight/Shutterstock; p9 JimCummings/Shutterstock; p10-11 PatSPictures/Shutterstock; p13 (inset) https://commons.wikimedia.org/wiki/File:Fur_trapper_in_Mount_St_Helen_area.jpg; p13 (main) Medstock/Shutterstock; p14-15 SusannePommer/Shutterstock; p17 ABPhotographia/Shutterstock; p19 https://commons.wikimedia.org/wiki/File:Doug_Smith,_senior_wildlife_biologist,_examines_the_teeth_of_a_wolf-killed_bull_elk_(d44734e0-8551-4ad3-bca4-c088aba0681d).jpg; p20-21 DanitaDelmont/Shutterstock; p22 Scigelova/Shutterstock; p23 ChaseDekker/Shutterstock; p25 MennoSchaefer/Shutterstock; p26-27 HarryBaugelink/Shutterstock; p29 BaiterekMedia/Shutterstock

Printed in the United States of America

CPSIA compliance information: Batch #CSGS23: For further information contact Gareth Stevens, New York, New York at 1-800-542-2595.

CONTENTS

Words in the glossary appear in **bold** type the first time they are used in the text.

A HUNTER IS HUNTED

For much of its history, the gray wolf roamed across North America as a fierce predator. At the top of the food chain, the gray wolf had no natural predators. It lived in peace with the Native Peoples of North America, who treated it with respect.

However, the wolf was no match for European settlement. Seeing the gray wolf as a threat, settlers killed off any wolves that got in their way. They harmed the gray wolf's **excosystem** and made it their mission to wipe out the gray wolf population for good. Eventually, the gray wolf was driven from its lands—almost to **extinction**.

A LEGENDARY CREATURE

NATIVE AMERICANS CREATED STORIES AND LEGENDS ABOUT THIS HOWLING HUNTER. SOME WORSHIPPED THE WOLF AND BELIEVED IT TO BE A CREATURE OF GREAT SPIRITUAL POWER. MANY **CULTURES** SAW THE WOLF AS GOD-LIKE, AND DIDN'T ALLOW FOR IT TO BE KILLED. MANY WATCHED THE WOLF FOR SURVIVAL CLUES, BELIEVING IT TO HAVE GREAT WISDOM.

European settlers and, later, U.S. citizens nearly hunted the gray wolf to extinction by the early 1900s.

THE GRAY WOLF UP CLOSE

Of all the wolf species, or kinds, gray wolves are the largest. They're also the largest members of the canine, or dog, family. They weigh between 31 to 143 pounds (14 to 65 kg). Gray wolves can have gray, black, white, or brown hair. Their fur color can help them blend in with their surroundings when following prey.

Wolf packs are usually very close family units. They live and hunt together. The alpha couple is the most powerful male and female of the pack. The members of a pack live together and help raise young. When the wolf pack hunts, each member has a certain job.

WOLF TALK

GRAY WOLVES ARE KNOWN FOR THEIR NIGHTTIME HOWLS. THAT SOUND HAS BECOME PART OF MANY SCARY STORIES. IN REALITY, THE WOLVES ARE JUST COMMUNICATING WITH ONE ANOTHER. THEY ALSO SPEAK THROUGH BODY LANGUAGE. FOR EXAMPLE, A PLAYFUL WOLF WILL BOW AND PRANCE AROUND, OR WALK IN A LIVELY WAY. AN ANGRY WOLF WILL GROWL AND SHOW ITS TEETH.

Scientists believe the dogs we keep as pets today **descended** from wolves many thousands of years ago. Do they look alike to you?

A "BEFORE" PICTURE

Gray wolves once lived unchallenged in North America. Their amazing senses of smell and hearing made them great hunters. Gray wolves are the apex, or top, predators in their ecosystems. They had few threats to their safety and were found all over the Northern **Hemisphere.** They lived from Canada to Mexico, including in today's United States, and in parts of Europe and Asia.

The Navajo, Cheyenne, Chippewa, and Cherokee respect wolves and see them as powerful, spiritual beings. Some Native American nations are involved in efforts to restore wolf populations. In 2021, groups representing many tribes asked President Joe Biden to do more to protect wolves.

AMAZING ADAPTATIONS

WOLVES ARE VERY ADAPTABLE CREATURES, WHICH MEANS THEY CHANGE TO LIVE BETTER IN THEIR SURROUNDINGS. THEY CAN LIVE IN VERY COLD AREAS, VERY WARM AREAS, AND AREAS THAT ARE IN-BETWEEN. THEIR THICK FUR KEEPS THEM WARM AND DRY IN COLD PLACES. IN WARMER WEATHER, THEIR FUR IS LESS THICK AND THEY CAN PANT TO COOL DOWN.

Wolf populations were not at risk before white settlers arrived. This wolf totem pole was created by a First Nations community in what is today Canada.

WOLF ADAPTATIONS

THE SETTLERS' ARRIVAL

The 17th century saw the arrival of many European settlers to North America. Native Peoples had lived peacefully with nature and many hunted wild game instead of keeping animals. But settlers brought livestock that wolves still sometimes see as food even today. So, people started targeting wolves. However, studies today show wolf attacks don't have a major effect on the livestock industry.

More livestock are killed by coyotes and pet dogs than by wolves. There are many effective ways of protecting livestock without harming wolves. Today, **conservationists** encourage farmers and ranchers to use fences and guard dogs to protect their livestock without hurting predators.

THE BIG BAD WOLF

SOME SETTLERS WERE TRULY AFRAID OF WOLVES. WOLVES HAVE LARGE TEETH AND CAN HAVE YELLOW, ORANGE, HAZEL, OR GREEN EYES. SETTLERS BEGAN TO KILL WOLVES OUT OF ANGER AND FEAR. ALTHOUGH SETTLERS WERE AFRAID OF BEING ATTACKED BY WOLVES, THEY DIDN'T NEED TO BE. IN TRUTH, WOLVES FEAR PEOPLE AND STAY AWAY FROM THEM.

Other ways farmers can help prevent wolf attacks on livestock include using special light and sound machines to scare them away. Since wolves tend to avoid people, herders or range riders can also help guard livestock.

TRAPPING AND KILLING

As white settlers spread west, they stole land from Native Peoples and native animals—including the gray wolf. They set traps that used bait to draw wolves close. While some people wanted to keep their livestock safe, others killed wolves just for sport.

In the late 1870s, people started to poison the dead bodies of animals that wolves often ate. When the wolves ate the poisoned meat, they usually died. Wolves weren't the only animals that suffered when people used poisoned bait. Other **scavengers** such as foxes and bears also died. The effects on the ecosystem were huge.

PAID HUNTERS

BY THE LATE 1800S, PEOPLE WERE HIRED TO GET RID OF WOLVES AND WERE PAID FOR EACH WOLF THEY KILLED. SOME WOLF KILLERS HUNTED WOLVES WITH DOGS. OTHERS SHOT THEM WITH GUNS. BETWEEN DIRECT HUNTING AND POISONING PROGRAMS, WOLF POPULATIONS DROPPED QUICKLY IN NORTH AMERICA.

Wolf trappers, or "wolfers" kept wolf hides (skin and fur) as a trophy, or prize, and for money.

DRIVEN TO EXTINCTION

Federal and state governments paid wolf hunters to kill gray wolves—$20 to $50 for each one, which was a lot of money in the 1800s. Settlers and the U.S. government also harmed wolves by killing off bison and other large prey animals in the Southwest. Wolves were dying of hunger without their prey. Naturally, they turned to livestock as another food source, which made farmers and ranchers hate them even more.

People continued to hunt the gray wolf until the late 1900s. Wolf hunting programs lasted until 1965. By that point, these wolves were nearly extinct in the United States. By 1926, there were no more gray wolves left in Yellowstone National Park.

Yellowstone National Park is located mostly in Wyoming with sections in Montana and Idaho. It once held a large gray wolf population.

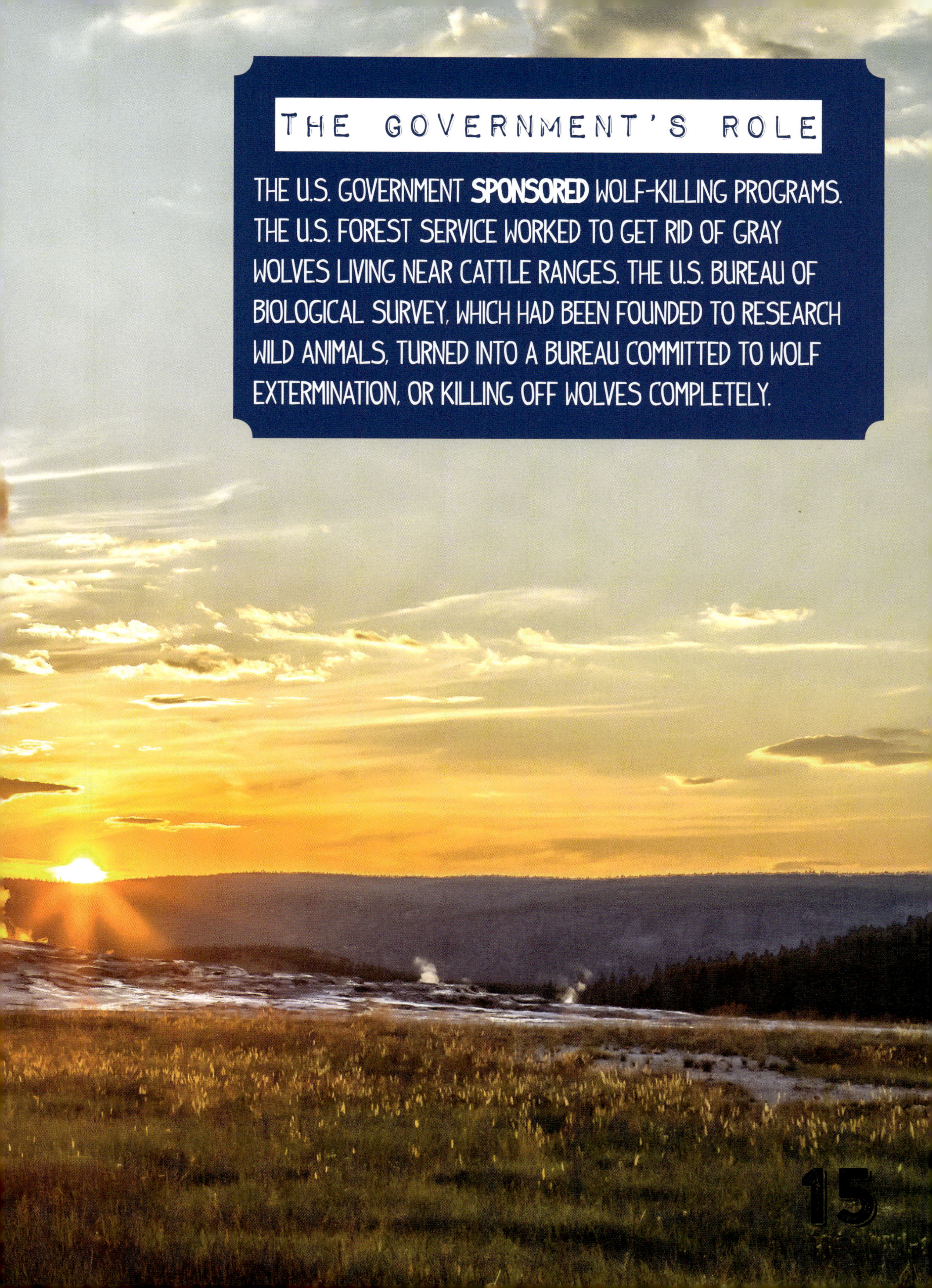

THE GOVERNMENT'S ROLE

THE U.S. GOVERNMENT **SPONSORED** WOLF-KILLING PROGRAMS. THE U.S. FOREST SERVICE WORKED TO GET RID OF GRAY WOLVES LIVING NEAR CATTLE RANGES. THE U.S. BUREAU OF BIOLOGICAL SURVEY, WHICH HAD BEEN FOUNDED TO RESEARCH WILD ANIMALS, TURNED INTO A BUREAU COMMITTED TO WOLF EXTERMINATION, OR KILLING OFF WOLVES COMPLETELY.

CONSERVATION NATION

Conservation efforts ramped up in the 1960s and 1970s. Wildlife experts, or authorities, educated the public about animals that were **endangered.** Some were endangered because of **systematic** killing, such as the gray wolf and American bison. Others were poisoned by chemicals used in farming called pesticides, such as the bald eagle.

On December 28, 1973, President Richard Nixon signed the Endangered Species Act. The act called for the conservation of endangered and threatened species. Under this act, the gray wolf became a protected species. People were no longer allowed to hunt the wolf legally, and a plan was put in place to help the species recover. This act marked the beginning of the gray wolf's recovery.

OTHER ENDANGERED ANIMALS

THE ENDANGERED SPECIES ACT HAD A HUGE EFFECT ON THE LIVES OF ANIMALS THAT WERE AT RISK OF DYING OUT. THE AMERICAN BISON, MANATEE, AND BALD EAGLE WERE ALSO BROUGHT BACK FROM THE BRINK OF EXTINCTION BECAUSE OF THE ENDANGERED SPECIES ACT. POPULATIONS WERE RESTORED THANKS TO ANIMALS BEING PROTECTED BY LAW.

By the late 1960s, more people began to realize the effects they had on nature and wildlife like wolves, especially thanks to the book *Silent Spring* by Rachel Carson.

THE YELLOWSTONE WOLF PROJECT

The ecosystem of Yellowstone National Park depends on the gray wolf. As a top predator, gray wolves affect the natural balance in the area. It was clear in the 1990s that wolves would have to be reintroduced into the park to bring back this balance.

The Yellowstone Wolf Project, led by biologist Doug Smith, sought to bring gray wolves back. In 1995 and 1996, 31 wolves were caught in Canada and released in Yellowstone National Park as part of the Yellowstone Wolf Project. In 1997, 10 more wolves from Montana were released there. Biologists observed the ecosystems to see how they would respond to the wolf's return.

APEX PREDATOR

GRAY WOLVES HAVE NO NATURAL PREDATORS OTHER THAN HUMANS. IN YELLOWSTONE, THEY HUNT BOTH LARGE AND SMALL PREY. THEY ARE ABLE TO TAKE DOWN HUGE BISON. IN THE WINTER, MOST OF THEIR DIET IS ELK. GRAY WOLVES CAN REACH SPEEDS OF AROUND 45 MILES PER HOUR (72.4 KM PER HOUR) WHEN THEY ARE CHASING DOWN THEIR PREY!

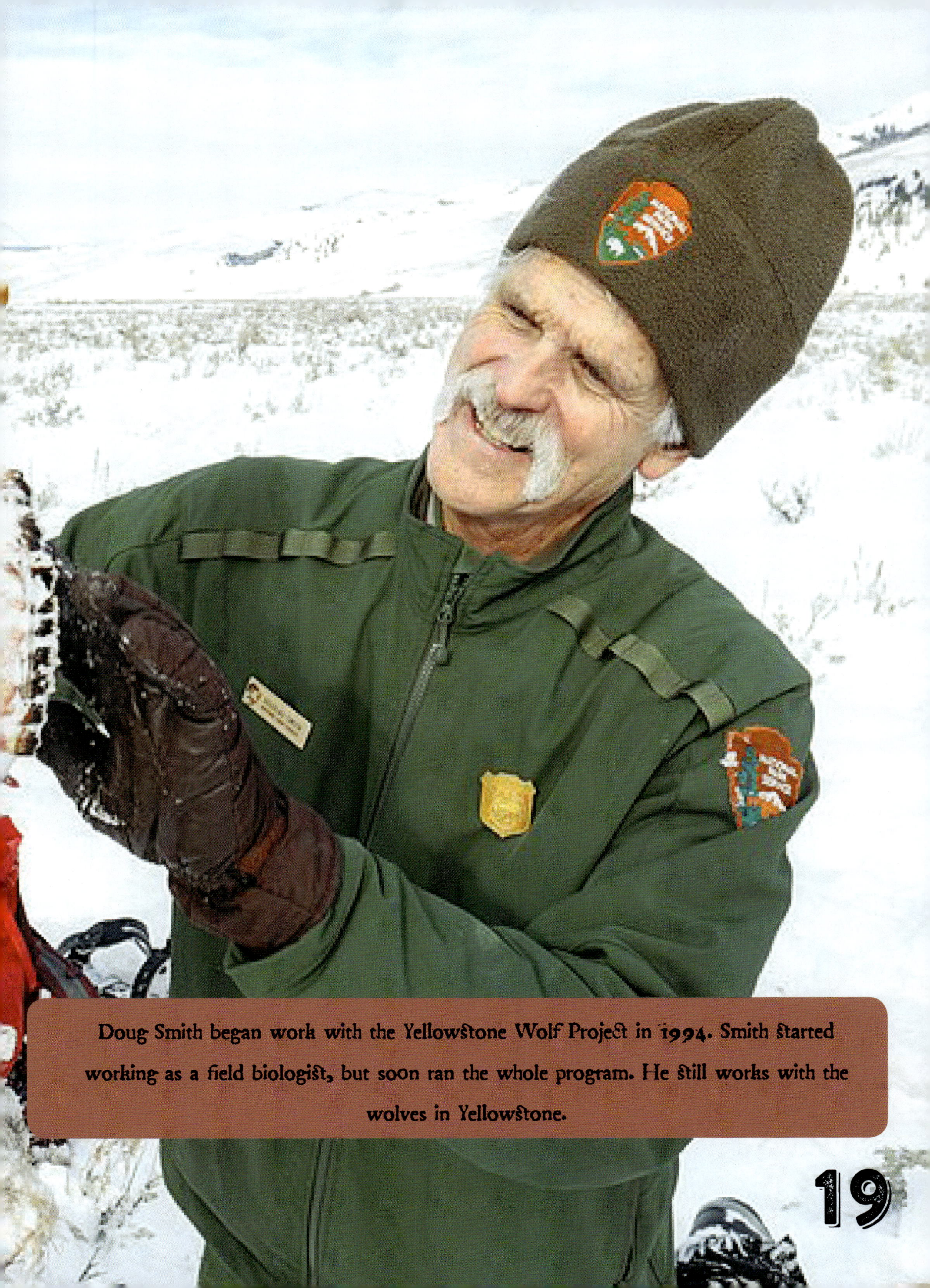

Doug Smith began work with the Yellowstone Wolf Project in 1994. Smith started working as a field biologist, but soon ran the whole program. He still works with the wolves in Yellowstone.

LEARNING ABOUT WOLVES

The Yellowstone Wolf Project keeps track of gray wolves with radio collars. The radio collars help scientists track their movements around the park. Doug Smith and other biologists involved with the Yellowstone Wolf Project catch wolves, **tranquilize** them, and put the collars on.

Smith and others collect data about the wolves' locations. The collars tell experts about wolf migration, or seasonal movement, and hunting styles. The tracking devices also tell biologists about individual wolf behaviors and life in a wolf pack. By tracking the wolves and other animals, the biologists have begun to understand the effect of the wolves on their ecosystem.

RAISING AWARENESS

LEAD BIOLOGIST DOUG SMITH CHARTED HIS FINDINGS WITH THE WOLVES OF THE YELLOWSTONE WOLF PROJECT. HE ALSO WROTE ABOUT HIS OBSERVATIONS IN JOURNALS. IN 2005, HE WROTE THE BOOK *DECADE OF THE WOLF: RETURNING THE WILD TO YELLOWSTONE*. HIS WRITINGS HELPED RAISE AWARENESS FOR THE YELLOWSTONE WOLF PROJECT.

From 2013 to 2019, around 20 percent of Yellowstone gray wolves were fitted with radio collars in June due to new pups being born.

WHAT HAPPENED TO THE ECOSYSTEM?

Each part of an ecosystem depends on other parts. If one animal is removed, others will grow out of control or suffer. The Yellowstone Wolf Project looked at what had happened to the ecosystem when wolves were killed off in Yellowstone.

When the wolves were gone, elk ate plants and trees in the same location all year round. They especially loved eating aspen, cottonwood, and willow trees. They would nibble on the young plants and the trees could never grow into adulthood. As elk damaged the population of trees, the habitat changed greatly. Habitats are the natural homes for plants and animals. Streams suffered from **erosion.** Beaver populations fell greatly.

WHAT DO BEAVERS HAVE TO DO WITH IT?

YELLOWSTONE BEAVERS RELY ON WILLOW TREES TO SURVIVE. WHEN ELK ATE THE TREES, THE BEAVERS LOST THEIR FOOD SOURCE. FEWER BEAVERS SURVIVED TO BUILD HOMES IN THE STREAMS, WHICH NORMALLY PROVIDE COLD, SHADED WATER FOR FISH. WITHOUT COLD, SHADED WATER, FISH STARTED TO DIE TOO.

It may be hard to see the connection between wolves and beavers, but in an ecosystem, all things are connected.

SEEING THE BENEFITS

After the wolf introduction, biologists started to see new changes in the Yellowstone ecosystem. The elk began to move around again as wolves hunted them. Deep snow was once the greatest threat to elk, but now wolves are the main cause of death in elk populations—as they should be.

Without as many elk, plants had time to recover and grow. This prevented erosion and provided shelter for smaller animals. Because more willow trees grew, the beaver population also increased from one colony in 1995 to nine in 2021. Scavengers also benefited from wolves killing elk. Bears, coyotes, eagles, and ravens could eat the wolves' leftovers.

A CHANGING MINDSET

DURING THE 1800S, WHEN WOLVES WERE SYSTEMATICALLY HUNTED AND KILLED, SETTLERS WEREN'T WORRIED ABOUT THE EFFECT IT WOULD HAVE ON THE ECOSYSTEM. THEY WERE ONLY CONCERNED WITH TAKING OVER THE LAND. PEOPLE NOW UNDERSTAND THAT LOSING ANY SPECIES, LARGE OR SMALL, IS HARMFUL TO AN ENTIRE ECOSYSTEM.

In 1997, only two years after gray wolves were brought back to Yellowstone, biologists were seeing positive changes in the ecosystem.

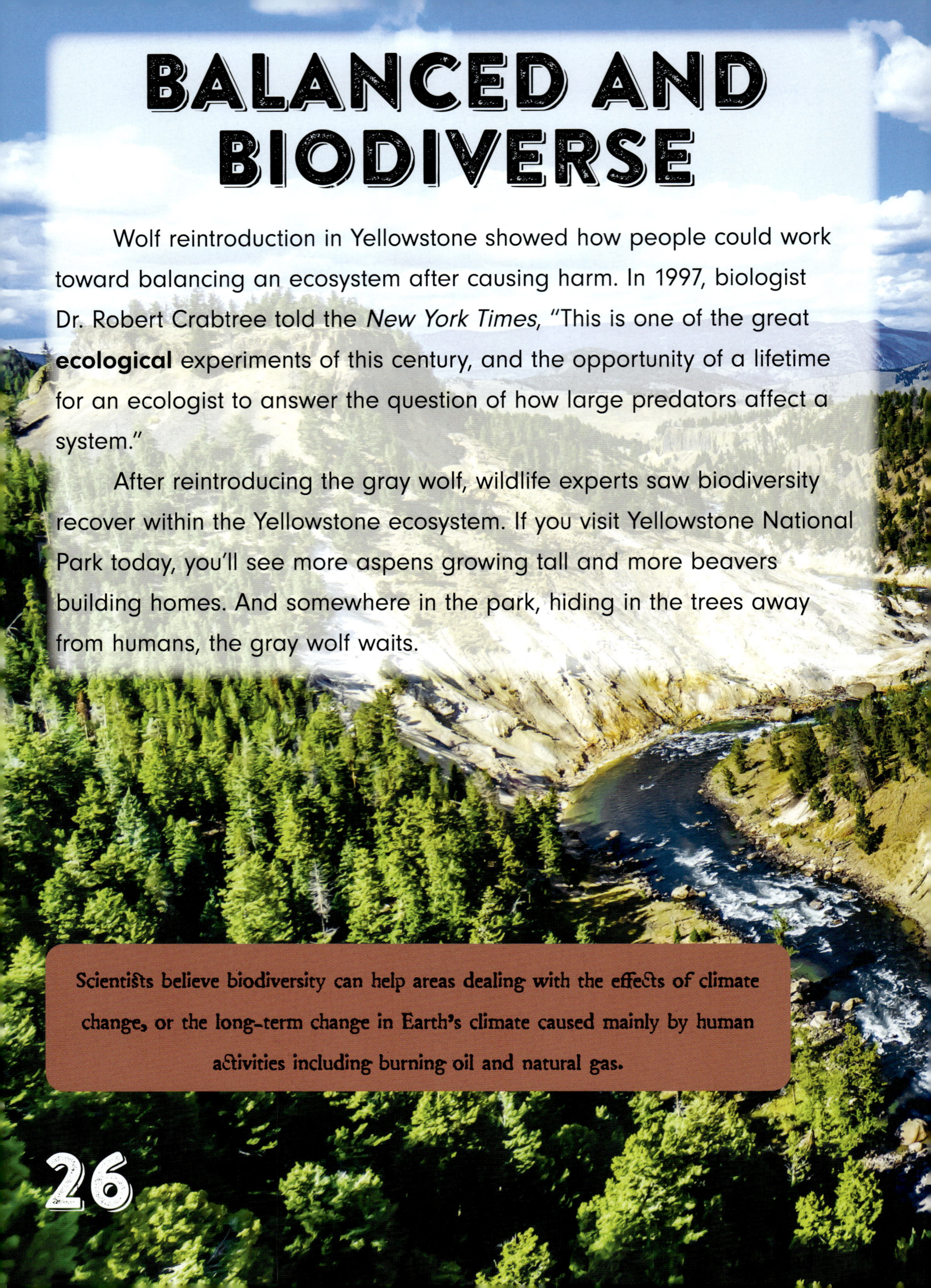

BALANCED AND BIODIVERSE

Wolf reintroduction in Yellowstone showed how people could work toward balancing an ecosystem after causing harm. In 1997, biologist Dr. Robert Crabtree told the *New York Times*, "This is one of the great **ecological** experiments of this century, and the opportunity of a lifetime for an ecologist to answer the question of how large predators affect a system."

After reintroducing the gray wolf, wildlife experts saw biodiversity recover within the Yellowstone ecosystem. If you visit Yellowstone National Park today, you'll see more aspens growing tall and more beavers building homes. And somewhere in the park, hiding in the trees away from humans, the gray wolf waits.

Scientists believe biodiversity can help areas dealing with the effects of climate change, or the long-term change in Earth's climate caused mainly by human activities including burning oil and natural gas.

A HEALTHY ECOSYSTEM

SCIENTISTS SAY THAT THE HEALTHIEST ECOSYSTEMS ARE THOSE THAT HAVE AS MANY NATIVE PLANTS AND ANIMALS AS POSSIBLE. THAT'S CALLED BIODIVERSITY. THE YELLOWSTONE NATIONAL PARK ECOSYSTEM LOST SOME OF ITS BIODIVERSITY WHEN THE GRAY WOLF WAS KILLED OFF. OTHER ECOSYSTEMS ACROSS THE COUNTRY MAY LOSE THEIR BIODIVERSITY BECAUSE OF HABITAT LOSS, CLIMATE CHANGE, AND OTHER RESULTS OF HUMAN ACTIONS.

DOING YOUR PART

The Yellowstone Wolf Project was a major undertaking for conservationists. The project showed that ecosystems can bounce back if they're given the chance. The work continues today. You can donate to or raise awareness of the Yellowstone Wolf Project. With continued support, the project can keep up its efforts to save and understand the wolves of Yellowstone National Park.

You can also tell others about the gray wolf. Give them facts about real wolf behavior. Help people understand that wolves aren't dangerous to people—and that they're necessary parts of their ecosystem. You can do your part to keep this wonderful wolf around for many years.

NOT THE VILLAN

WOLVES AREN'T THE BLOODTHIRSTY **VILLANS** WE SEE IN MOVIES AND ON TELEVISION. THEY JUST WANT TO LIVE AND HUNT IN PEACE. IN FACT, THERE HAVE ONLY BEEN TWO CASES OF A WOLF KILLING A PERSON IN NORTH AMERICA RECORDED SINCE 1900. IF PEOPLE STOP FEARING THE GRAY WOLF, WE CAN WORK EVEN HARDER TO PROTECT IT.

GREATER YELLOWSTONE ECOSYSTEM WOLF POPULATION (1995–2015)

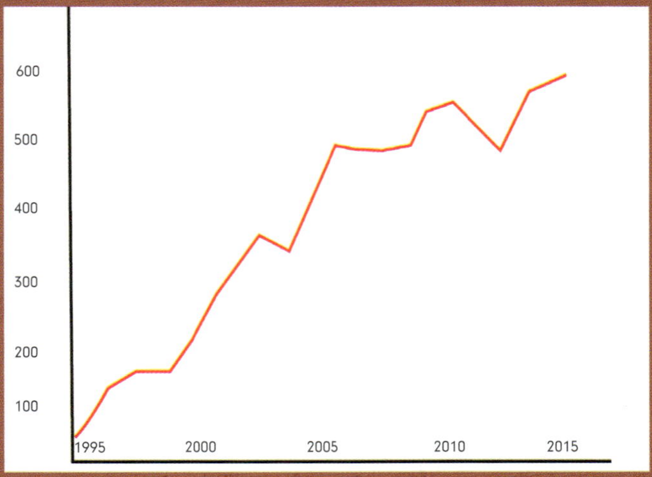

Without the Yellowstone Wolf Project, this great hunter wouldn't be roaming Yellowstone National Park today.

GLOSSARY

conservationist: Someone who works to protect nature.

culture: The beliefs and ways of life of a group of people.

descended: Born of a certain group or family.

ecological: Having to do with the study of how living things affect one another.

ecosystem: A natural community of living and nonliving things.

endangered: In danger of dying out.

erosion: The wearing away of Earth's surface by wind or water.

extinction: The death of all members of a plant or animal species.

hemisphere: One-half of Earth.

scavenger: An animal that eats what it can find, including waste and dead animals.

sponsor: To pay for the cost of an activity, organization, or event.

systematic: Using a careful system or method to do something.

tranquilize: To use a drug to cause an animal to become very relaxed and calm.

villain: Someone or something that is blamed for a problem.

FOR MORE INFORMATION

Books

Ardely, Anthony. *What If Gray Wolves Disappeared?* New York, NY: Gareth Stevens Publishing, 2023.

Barr, Catherine. *Fourteen Wolves.* London, UK: Bloomsbury Children's Books, 2021.

Borgert-Spaniol, Megan. *Gray Wolves: Yellowstone's Hunters.* Minneapolis, MN: Super Sandcastle, 2020.

Websites

Gray Wolf
naturemappingfoundation.org/natmap/facts/gray_wolf_k6.html
Discover facts about the gray wolf—and where you can find it.

Yellowstone National Park
kids.nationalgeographic.com/nature/article/yellowstone
Explore Yellowstone National Park with National Geographic Kids and learn more about the wildlife there.

Yellowstone Wolf Project
www.yellowstone.org/wolf-project/
Learn more about the Yellowstone Wolf Project and how it works to keep gray wolf populations stable.

INDEX